ONAP Demystified

Automate Network Services with ONAP

Amar Kapadia

ONAP Demystified

ISBN: 1720879915
ISBN-13: 978-1720879916

Copyright © 2018 The Linux Foundation under license to Aarna Networks, Inc.

All rights reserved. No part of the book may be reproduced, stored in a retrieval system, or transmitted in any form or by any means without the prior written permission of the publisher, except in the case of brief quotations embedded in critical articles or reviews.

Every effort has been made in the preparation of this book to ensure the accuracy of the information presented. However, the information contained in this book is sold without warranty, either express or implied. Neither the authors nor Aarna Networks, Inc. and its dealers and distributors will be held liable for any damages caused or alleged to be caused directly or indirectly by this book.

Aarna Networks, Inc. has endeavored to provide trademark information about all the companies and products mentioned in this book by the appropriate use of appropriate marks. However, Aarna Networks, Inc. cannot guarantee the accuracy of this information.

First publication:	June 2018
Second publication:	August 2018
Third publication:	December 2018

Published by Aarna Networks, Inc.
2670 South White Road, Suite 254
San Jose, CA 95148
www.aarnanetworks.com

CREDITS

AUTHOR

Amar Kapadia

REVIEWERS

Amir Levy

Flavia Cioanca

Lisa Caywood

Eric Debeau

Magda Stepien

Michael Still

Ranny Haiby

Sundar TJ

COVER WORK

Liu Zishan

GRAPHICS

Amar Kapadia

PROOFREADER

Heaven Hodges

ACKNOWLEDGEMENT

I would like to thank my wife Aditi Kapadia, who has provided constant encouragement and patiently tolerated my weekend and late-night book-writing sessions. I would also like to thank the ONAP community members who painstakingly reviewed the content and The Linux Foundation staff that helped with the content.

CONTENTS

PREFACE	9
What This Book Covers	9
1 NFV BASICS & INTRODUCTION TO ONAP	11
Introduction	11
Why NFV?	11
Network Services Before NFV	12
What is NFV?	13
What is SDN?	14
Benefits of NFV	14
Case Study British Telecom	15
Case Study Telstra PEN	15
Case Study Orange	15
NFV Use Cases	16
The Role of ETSI in NFV	17
ETSI NFV Architecture	17
Impact of NFV Transformation on the Organization	19
NFV Transformation Steps	20
TIP People & Processes	21
Open Source and NFV Transformation	21
Why ONAP?	23
Conclusion	24

2 ONAP SCOPE & KEY CONCEPTS — 25
- Introduction — 25
- ETSI MANO Scope — 25
- What is ONAP? — 26
- The History of ONAP — 27
- ONAP Scope — 27
- ONAP and Other Software Systems — 28
- ONAP Primary Architectural Principles — 29
- Model Driven — 30
- What Gets Modeled? — 31
- Modeling Languages — 32
- TOSCA — 32
- YANG — 33
- Other Modeling Languages — 34
- Cloud Native Design Principles — 35
- DevOps — 35
- Other Architectural Principles — 36
- Day in the Life of ONAP — 37
- ONAP Releases — 37
- Conclusion — 38

3 ONAP ARCHITECTURE — 39
- Introduction — 39
- ONAP Architecture — 39
- ONAP Design Time Environment — 41
- VNF Onboarding Process — 42

VNF Onboarding Projects	43
ONAP OSS/BSS Interface	44
ONAP as a Whole	45
ONAP Enables Agility	46
ONAP Implementation	47
Conclusion	47
4 ONAP DEEPDIVE	**49**
Introduction	49
Official ONAP Projects	50
ONAP Lifecycle Management	50
ONAP Design Time: SDC	51
CLAMP	52
SO	52
ONAP Run Time: Controllers	53
APP-C	54
VF-C	54
SDN-C	54
MultiCloud (or MultiVIM)	54
DCAE	55
Policy	55
A&AI	56
ONAP Portal	57
ONAP Northbound APIs	58
Other Design Time and Run Time Projects	58
ONAP Common Services	59

ONAP Supporting Projects	61
ONAP Project Structure	62
ONAP Releases	63
Conclusion	64
5 ONAP BLUEPRINTS & GETTING INVOLVED	**65**
Introduction	65
vFW Blueprint	65
Residential vCPE Blueprint	67
vCPE Technical Walkthrough	67
Voice-over-LTE Blueprint	68
VoLTE Technical Walkthrough	69
CCVPN Blueprint	70
CCVPN Technical Walkthrough	71
5G Blueprint	72
Getting Involved	72
Conclusion	73
6 ADDITIONAL INFORMATION	**75**

PREFACE

Network functions virtualization, or NFV, is a once in a generation disruption that will completely transform how networks are built and operated. This mega-trend will affect telecom/cable operators and technology providers alike. 5G, edge computing and IoT all consider NFV a critical prerequisite.

Open source has revamped how enterprises build out their IT systems. Now, open source promises to do the same for NFV. The Linux Foundation Open Network Automation Platform (ONAP) project, according to the project website (onap.org), "provides a comprehensive platform for real-time, policy-driven orchestration and automation of physical and virtual network functions that will enable software, network, IT and cloud providers and developers to rapidly automate new services and support complete lifecycle management." NFV adoption has been slower than just about anybody expected, and the lack of a robust, well-architected and broadly adopted automation platform has been one of the inhibitors. ONAP fills this gap. In the next six chapters, you will find out what ONAP is and why you should care.

What This Book Covers

Chapter 1, NFV Basics & Introduction to ONAP, introduces NFV, elements of NFV transformation and the motivation for ONAP. By reading this chapter, you will be able to articulate what NFV is and its benefits, provide a high-level explanation of the ETSI NFV architecture, discuss the elements of NFV transformation and understand the gaps ONAP is filling.

Chapter 2, ONAP Scope & Key Concepts, looks at the scope of the ONAP project and key architectural concepts. With this chapter, you will be able to state the scope of the ONAP project, understand the systems ONAP interacts with, identify the key architectural concepts behind ONAP and understand the roles of different individuals that will use ONAP.

Chapter 3, ONAP Architecture, covers the ONAP architecture and how the various components work together to accomplish network automation. By the end of this chapter, you will be able to discuss the ONAP internal high-level architecture, describe the different components that make up the design time modules and their roles, describe the different components that make up the runtime modules and their roles, articulate the three major interfaces and characterize how the different ONAP modules work together.

Chapter 4, ONAP Deep dive, covers all the sub-projects that make up ONAP. Will also look at the project organization. By reading this chapter, you will be able to understand the ONAP block diagram, explain how ONAP is deployed and managed, articulate the various design time projects, describe the run time projects, discuss the Portal project and ONAP APIs and list common services and supporting projects.

Chapter 5, ONAP Blueprints & Getting Involved, discusses the official demos, or blueprints, that the ONAP community has created. In this chapter, you will learn about the vFW, vCPE, VoLTE, CCVPN, and 5G blueprints. You will also learn how to get involved with the project.

Chapter 6, Additional Information, lists useful links.

1

NFV BASICS & INTRODUCTION TO ONAP

Introduction

In this chapter, we will get an introduction to NFV, elements of NFV transformation and the motivation for ONAP.

Why NFV?

Communications service providers (CSPs) are facing competition from over-the-top and web services, experiencing declining average revenue per user (ARPU) and feeling the pressure to innovate rapidly to respond to new trends such as 5G, IoT and edge computing. Network Functions Virtualization (NFV) is a technology that can greatly assist in solving these business challenges.

Network Services Before NFV

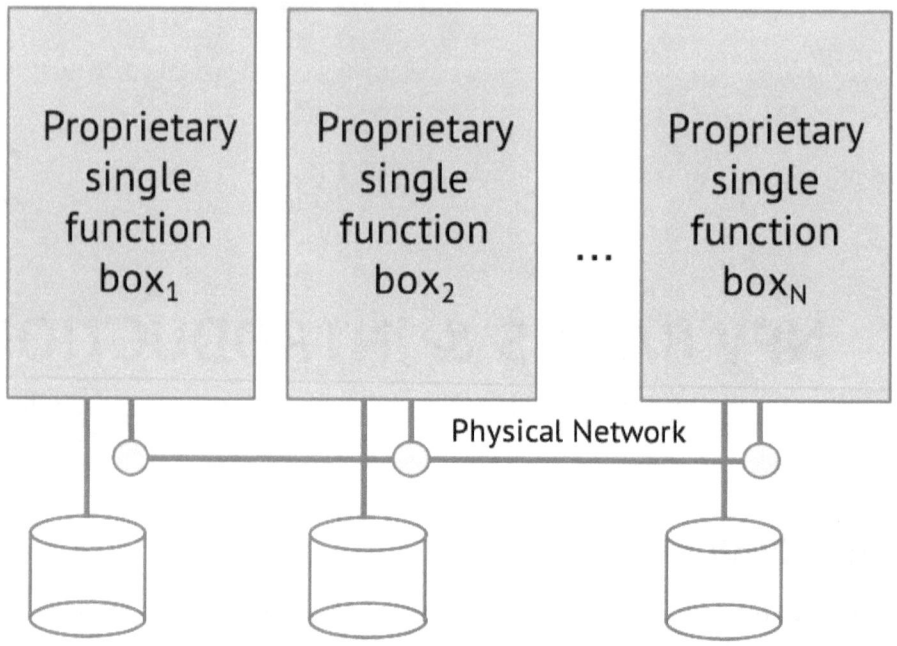

Dedicated local or network storage

Network Services Before NFV

Traditional network services are built by chaining together proprietary single function boxes. The design of these services is custom, and the underlying equipment is expensive, requires lengthy procurement times and cannot be shared with any other service. Once deployed, the operations and management of these services are largely non-automated with each box presenting its own management interface. This technique for creating network services is very expensive and offers no practical way to create dynamic services.

What is NFV?

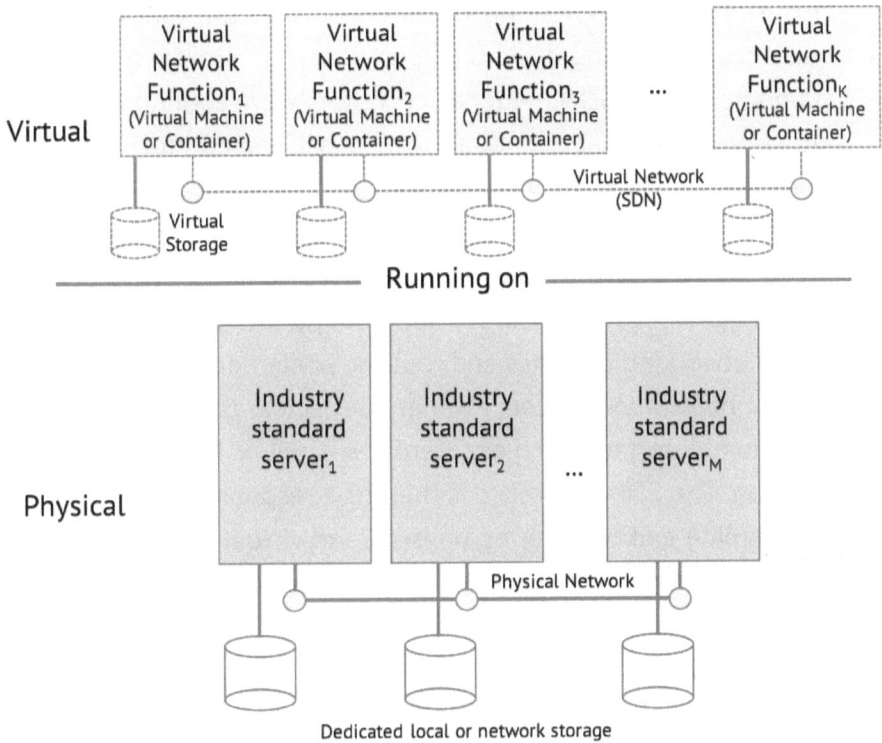

Network Services Using NFV

NFV virtualizes each single function box used in current network services. Once virtualized, the virtual network function (VNF) can be hosted on an industry-standard server. Virtualization does not stop at replacing physical boxes with virtual machines but can go further via microservices, containers and cloud-native architectures. Lifecycle operations such as initial deployment, configuration changes, upgrades, scale-out, scale-in, self-healing etc. are all simply software steps that can be fully automated. These VNFs can also be chained together and managed in a dynamic and automated fashion. NFV thus results in agile on-demand services.

What is SDN?

Software-defined networking, or SDN, separates the control plane of traditional networking devices (e.g., switches and routers) from the data plane. Before SDN, switches or routers were configured via routing protocols that did not allow fine-grained control. The control plane is known as an SDN controller.

SDN controllers have a northbound interface that connects with cloud and management systems and a southbound interface to physical and virtual networking switches and routers. While northbound interfaces have not been standardized, a variety of standards exist for the southbound interface, such as OpenFlow, OpFlex, Netconf, P4, ovsdb, and so on. The SDN controller is therefore responsible for programming and configuring physical and virtual networking elements (layers 2 and 3). SDN controllers can also be used to program and configure layers 4–7.

Throughout the rest of the book, when we say NFV, the student can assume we mean both SDN and NFV.

Benefits of NFV

NFV provides four primary benefits to CSPs:
1. Increased revenue: The combination of introducing new services faster and existing services in a more dynamic fashion can jointly result in increased revenue.
2. Reduced capital expenditures: The use of industry-standard servers and switches, increased hardware utilization and the adoption of open source software results in reduced capital expenditures.

3. Reduced operational expenditures: Automation and hardware standardization can substantially slash operational expenditures.
4. Improved customer satisfaction: The combination of service agility and self-service can result in greater customer satisfaction.

Case Study British Telecom

Intel and British Telecom jointly commissioned a study to understand the cost benefits of a particular NFV use case (virtual customer premises equipment or vCPE). Take a look at the study builders.intel.com/docs/networkbuilders/The-benefits-of-virtual-CPE-business-user.pdf. What percentage cost reduction can a large enterprise expect when using vCPE?

Case Study Telstra PEN

Telstra used to take 3 weeks to provision WAN bandwidth for customers. The process was manual and required sales people, program managers, network managers, purchase orders and meetings. The customer then had to commit to the bandwidth for one year. With their innovative PEN service, Telstra was able to cut down WAN bandwidth provisioning time to just seconds and customer commitment periods to just one hour. The process was a self-service and automated experience. Read more at telstraglobal.com/images/assets/partnerprogram/Connectivity/746_PEN_Short_V04.pdf.

Case Study Orange

Orange's first commercial SDN/NFV offering was the Easy Go Network in 2016. Provided by Orange Business Services, the Easy Go Network helps companies instantly provision branch connectivity by deploying virtual network functions (VNFs) on a universal customer premise equipment (uCPE) device. Ordering, customer care and reporting functions are all handled through a portal. Easy Go generates new revenue, cuts cost, improves customer satisfaction and speeds up service provisioning. Orange is also an active Linux Foundation Networking member, heavily involved in ONAP. See more at opendaylight.org/use-cases-and-users/user-stories/lf-networking-projects-power-next-generation-orange-networks.

NFV Use Cases

Here are some popular NFV use cases. The list is, by no means, comprehensive:

Virtual customer premise equipment (vCPE): This use case virtualizes popular enterprise and residential connectivity boxes such as firewall, router, SD-WAN, VPN, NAT, DHCP, IPS/IDS, PBX, transcoders, WAN optimization and set-top boxes etc. Once virtualized the VNFs can be hosted on a box at a customer's premises or in the CSP edge or cloud locations.

Virtual evolved packet core (vEPC): vEPC allows mobile network operators (MVNO) and enablers (MVNE) to use a virtual infrastructure to host their core mobile network.

Virtual IP multimedia system (vIMS): This use case allows a CSP to offer IP services (such as voice-over-IP) on their network. In combination with vEPC, CSPs can offer a voice-over-LTE (VoLTE) solution that cuts cost and improves customer experience.

Virtual radio area network (vRAN): With 5G on the horizon, vRAN allows a CSP to virtualize the radio area network and use commodity hardware to host the VNF.

The Role of ETSI in NFV

In 2012, ETSI took the reins of the standardization process for NFV with the goal of defining what NFV was and what the industry was going to do. The Industry Specification Group (ISG) for NFV at ETSI has grown from seven to more than 300 individual companies and has published more than 70 documents with details of how NFV should be implemented and tested.

ETSI NFV Architecture

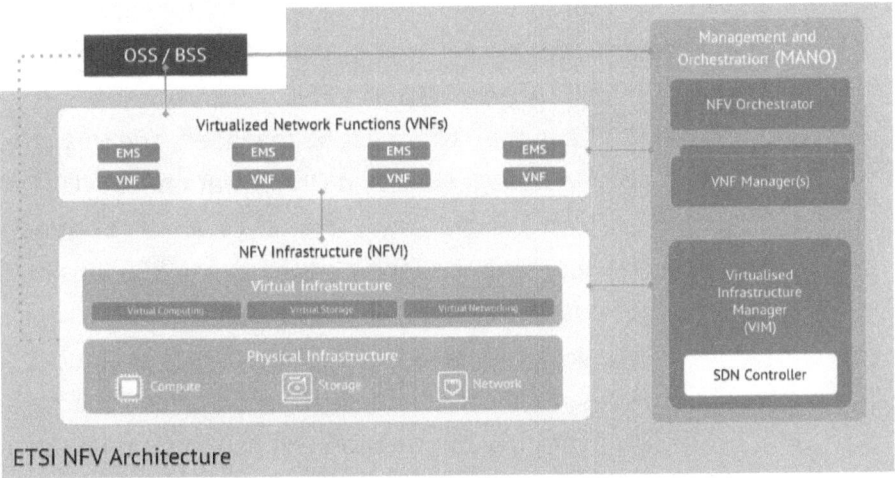

ETSI NFV Architecture

ETSI NFV Architecture

Perhaps the most important contribution of the ETSI NFV ISG was articulating a consistent reference architecture and interface definitions. The main components of the architecture are:

- The Network Functions Virtualization Infrastructure (NFVI) is the actual infrastructure on which NFV services are delivered. NFVI includes
    - Physical hardware such as servers, storage, network hardware
    - Virtual infrastructure, such as virtual compute (virtual machines or containers), virtual storage, and virtual networks (overlay networks).
    - Data plane acceleration technologies to compensate for the packet-processing performance loss experienced when going from physical to virtual
- One or more Virtual Network Functions (VNFs) run on top of the NFVI. Each VNF also comes with its element management system (EM); however, this fragmented approach is non-optimal, and there are moves in the industry to unify management into one entity (e.g., Linux Foundation initiatives such as ONAP DCAE, A-CORD and PNDA.io.).
- Both the NFVI and the VNFs must be managed, and this is handled by NFV Management and Orchestration (MANO). NFV MANO doesn't act in isolation; it interacts with the existing external OSS (operations support systems) and BSS (business support systems) components. MANO includes
    - Virtualized infrastructure manager (VIM) or cloud management software (e.g., OpenStack or Kubernetes).
    - SDN controller to provision and configure the underlay (physical) and overlay (virtual) networks. The SDN controller can also be used to configure VNFs.
    - VNF manager that manages the lifecycle of one VNF at a time.
    - NFV orchestrator that manages the lifecycle of an entire network service (which is constructed by multiple VNFs chained together).

Impact of NFV Transformation on the Organization

NFV transformation does not affect just a technology; instead, it impacts many aspects of your organization:

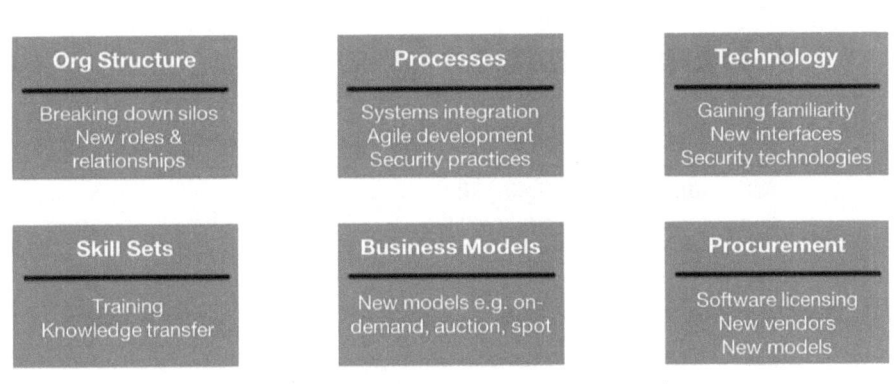

NFV Drives Organizational Transformation

The six areas of the organization impacted are:

- **Organization structure:** Small autonomous teams are required to enable self-contained projects.
- **Process:** Waterfall project management techniques need to be replaced with Agile development to increase delivery velocity.
- **Technology:** As we will cover in the next chapter, model-driven architectures, cloud-native design and DevOps need to be embraced.
- **Skill sets:** These new technologies require new skill sets. These skills will have to be inculcated organically because there just aren't enough external people available.
- **Business models:** Agile services and new trends like network slicing will require CSPs to innovate on business models as well.
- **Procurement:** The disaggregation of the network stack and the use of open source software will require soul-searching on the

part of CSPs. While the entire technology stack was outsourced to network equipment manufacturers (NEMs in the past), in this new era, a CSP can use the same outsourcing approach, take over the entire development and integration burden inhouse, or do something in-between.

NFV Transformation Steps

1 Articulate goals	2 Build skills organically	3 Agile development	4 Small steps
5 Clear use case	6 Executive sponsor	7 Dedicated team	8 Spread knowledge

NFV Transformation Steps

Some of the best practices of other CSPs, such as KPN and Deutsche Telekom are:

1. **Clearly articulate goals:** It is important to pick the right goals and communicate them broadly. For example, a goal could be reducing service deployment time from weeks to minutes.
2. **Build skills organically:** As we discussed, there just aren't enough skilled resources outside, so there needs to be an emphasis on internal skill-building.
3. **Agile development:** Instead of adopting a massive initiative upfront, small projects need to be tried out, corrected and changed with time. See how Orange chose a relatively narrow, well-contained use case (Easy Go) as one of their early NFV initiatives.

4. **Small steps:** There is no need to jump to complete network automation with the first project. Smaller steps reduce risk and improve the chances of success.
5. **Clear use case:** Rather than building an NFV cloud and waiting for use cases to show up, a use case needs to be identified up front.
6. **Executive sponsor:** Large initiatives, such as NFV transformation, need to be driven both top-down and bottom-up. Therefore, an executive sponsor is critical for the success of the project.
7. **Use dedicated teams:** It is tempting to burden an existing team with NFV. However, a new, dedicated team that embraces a new culture, set of tools and processes is more likely to be successful.
8. **Spread the knowledge:** The initial team needs to act like a catalyst that transfers its newly acquired skills, knowledge, thinking, culture and processes to the rest of the organization.

TIP People & Processes

Another resource that helps with NFV transformation is the People & Process (P&P) subproject under the Telecom Infra Project (TIP).

Open Source and NFV Transformation

Open source projects have touched every aspect of enterprise IT. The CSP space is no different. There is a plethora of open source projects addressing NFV requirements. Open source software serves as a supplement to standards to promote interoperability. Additionally, open source software provides:

- Improved interoperability through open APIs
- Higher Innovation velocity

- Faster troubleshooting capability
- Greater ability to influence roadmaps by direct R&D investment
- Reduced vendor lock-in, resulting in lower long-term costs
- Transparency
- Community support
- Larger talent pool

Why MANO (Management and Orchestration)?

Compute, storage and network virtualization

Elastic, on-demand, API driven NFV infra

Automated Ops, Admin and Mgmt. (OAM)

NFV is Not Just Virtualization

Though a common misconception, virtualization is not the same as NFV. The first step of NFV is to virtualize the infrastructure. The second step is to wrap the infrastructure with a cloud layer that provides API driven, elastic, self-service virtual infrastructure. Note that it is possible to combine the first and second steps by using a private cloud platform like OpenStack to provide a cloud layer with virtualization "out of the box." The third step is to fully automate OAM (operations, administration and management). This means that the orchestration and lifecycle management of VNFs and network services (that are built by chaining together VNFs) is automated. It also means that monitoring and service assurance are automated and architected to drive the lifecycle management, resulting in full network automation. For this reason, the MANO[1] component of the ETSI NFV architecture is critical for the success of NFV.

Why ONAP?

Many open source and proprietary MANO systems have existed for years. So why was ONAP required? Here are some of the reasons:
- Incomplete or old-school solutions did not meet the requirements:
    - Automation requires design, network service lifecycle management and service assurance capabilities — all working together
    - Agility requires innovative architectural approaches (see next chapter)
    - Automation further requires the ability to create analytics and AI/ML (artificial intelligence/machine learning) applications
    - Upcoming 5G and IOT trends require the ability to serve edge and core use cases, including multi-access edge computing (MEC) applications
- Prior products/projects did not have enough momentum to drive:
    - Standardized VNF onboarding mechanisms along with testing and validation
    - Cloud-native VNFs
    - Network service lifecycle design and management
    - Service assurance
    - Standardization of model driven languages
- Open source projects have an ability to execute a code-first approach from an API and interface specification point of view.

---

[1] From here on, we will mean NFVO and VNFM when we use the term MANO; and will exclude VIM and SDN controller.

This contrasts with standardization efforts, which may be restrictive and may not be able to keep up with the rapid pace of innovation. In other words, early on, an open source method of creating standards may be more efficient than that of traditional standard bodies.

These are some of the reasons the ONAP project was created.

Conclusion

The next-generation network is being built today, and it's being built with open source tools. We reviewed what NFV and SDN are, their benefits and their use cases. We also looked at the NFV architecture and what it takes for an organization to undergo NFV transformation. Finally, we looked at why ONAP is required.

2

ONAP SCOPE & KEY CONCEPTS

Introduction

In this chapter, we will look at the scope of the Linux Foundation Open Network Automation Platform (ONAP) project and the key architectural concepts. To drive complete automation, ONAP approaches things differently than prior solutions. For this reason, it is important to fully understand the various architectural concepts.

ETSI MANO Scope

Let us first revisit the scope of the MANO software component (again we are restricting the definition of MANO to include only NFVO and VNFM, not VIM):

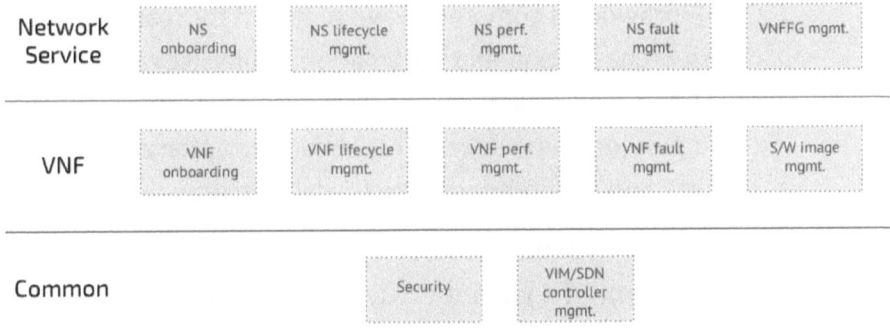

ETSI MANO Scope

Like ETSI, MEF (an industry association of 200+ companies enabling agile, assured, and orchestrated communication services that empower users with the dynamic performance and security required to thrive in the digital economy) is actively involved in service orchestration. Its MEF3.0 Lifecycle Service Orchestration (LSO) effort places special emphasis on orchestration across multiple service providers and multiple network technology domains. Additionally, the TM Forum (global industry association that drives collaboration and collective problem-solving among communication and digital service providers and their ecosystem of suppliers) actively works in the service orchestration area with its ZOOM project (Zero-touch Orchestration, Operations and Management).

What is ONAP?

ONAP is an open source Linux Foundation project. According to the official website:

> ONAP provides a comprehensive platform for real-time, policy-driven orchestration and automation of physical and virtual network functions that will enable software, network, IT and cloud providers and developers to rapidly automate new services and support complete lifecycle management.
>
> By unifying member resources, ONAP will accelerate the development of a vibrant ecosystem around a globally shared architecture and implementation for network automation –

with an open standard focus – faster than any one product could on its own.

Given the ETSI context, ONAP may be considered a MANO++ project, that is software that covers the entire scope of ETSI MANO and goes beyond it. ONAP is part of the Linux Foundation Networking Fund (LFN) and enjoys broad participation from both CSPs and vendors.

The History of ONAP

ONAP resulted from the merger of two open source projects: OpenECOMP and Open-O. AT&T had an internal network automation project called Enhanced Control, Orchestration, Management & Policy (ECOMP) that was open sourced under the Linux Foundation in late 2016/early 2017 in collaboration with Orange. ECOMP was already in production at AT&T when it was open sourced under the name OpenECOMP. Open-O was an earlier Linux Foundation MANO project. In early 2017, OpenECOMP and Open-O merged to form ONAP.

ONAP Scope

ONAP Scope

The scope of ONAP includes a comprehensive design framework to create network services and related items (covered later) and a runtime framework. The delineation is useful since it allows clean partitioning of work between product managers, designers, testers and operations. The runtime framework includes the ETSI MANO functionality of NFVO and VNFM but goes beyond that by including monitoring and service assurance. The two runtime blocks work in tandem to create closed-loop automation, where an event can trigger a corrective lifecycle management event without any human intervention. The policy framework, along with closed loop automation, plays a key role in the journey to full automation.

ONAP and Other Software Systems

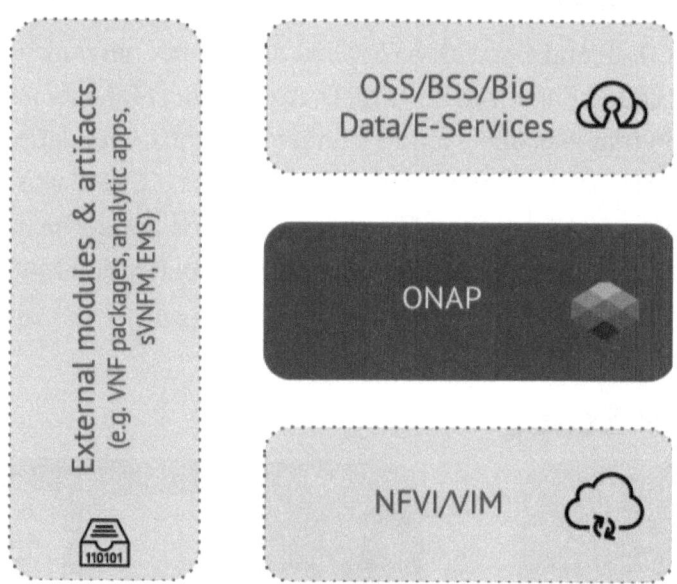

ONAP and Other Software Systems

ONAP primarily interacts with these software systems:
- **Operational support systems (OSS):** OAM, FCAPS (fault, configuration, accounting, performance, security)
- **Business support systems (BSS):** Service provisioning, deployment, service change management, customer information management, SLA management, billing and customer support

- **Big data applications:** Analytics applications to gain business insights and intelligence from data lakes containing network data collected by ONAP
- **E-services:** Self-service portals where customers can manage their services; essentially a BSS

ONAP Primary Architectural Principles

ONAP approaches things differently than traditional software systems. Here is my interpretation of the top three ONAP architectural principles.

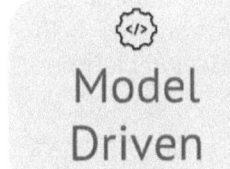

Automated without hard-coding

Built for the cloud and to manage cloud native VNFs

Built using CI/CD manage VNFs using CI/CD

ONAP Architectural Principles

Let's look at each principle in more detail.

Model Driven

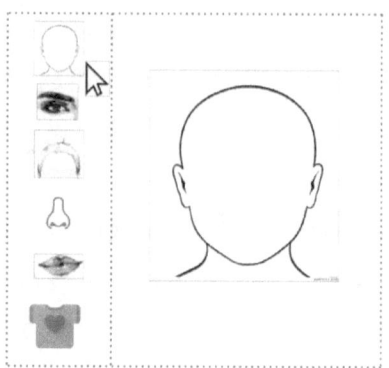

Custom: service dependent
Manual
Different each time

Standardized: service independent
Automated
Identical each time

What is Model Driven?

In the era before model-driven, network services would be built like the left-hand side image. Each network service would be custom designed. It would be manually built and hand-crafted. It would be dedicated; that is, the implementation would only support that one service. Finally, every time a similar service was built, it would be different.

Automation can be achieved in two ways. First, a bunch of programmers could write scripts and programs to automate the creation and deployment of a particular service. While this is great, it creates problems in terms of requiring skilled programmers. Making changes is also very difficult. Also, this doesn't solve the custom, service-dependent and "different each time" issues we discussed above.

Models allow for automation without hardcoding. Similar to the right-hand side image, in model driven, network-service-independent artifacts are defined using a human readable modeling language and are stored in a catalog. These models are put together to create network services. Since there's no code involved, you don't need an army of programmers. Services can also be changed easily by changing models. Finally, the service is deployed in exactly the same way each time as described by the model. Needless to say, the modeling language and the model itself both have to be standardized.

What Gets Modeled?

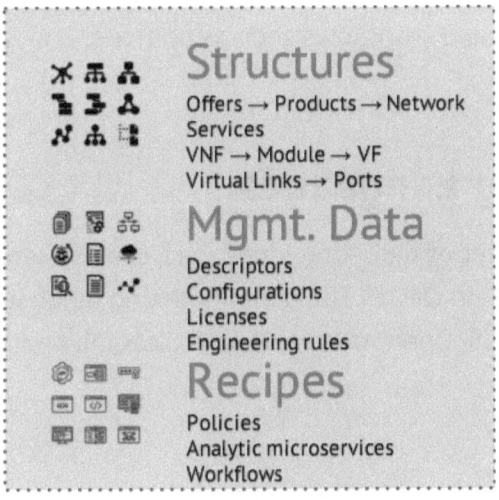

What Gets Modeled

A wide variety of artifacts get modeled in ONAP. The first set of things that get modeled consist of structures. At the highest level, there are offers. Offers consist of products and marketing configurations. Products consist of network services and commercial configurations, such as pricing. Network services consist of multiple VNFs, SDN services or both. VNFs consist of modules. Modules consist of virtual functions (VF), virtual links and ports. One VF maps to one virtual machine. SDN services specify underlay (physical) or overlay (virtual) network connectivity, often pertaining to inter-region or WAN connectivity.

The next set of items to get modeled is management data. This includes descriptors, licenses, configurations and engineering rules. Structures and engineering rules models can be considered empty shells; once populated they combine to form objects. Some examples of the above items are VNF descriptors, per CPU/perpetual licenses, VNF configurations and engineering rules requesting 4 vCPUs for a VF.

Additionally, ONAP model recipes can include policies, analytic microservices and workflows such as upgrade. Populated recipes result in functions.

Modeling Languages

ONAP uses a set of modeling languages, each suited for a specific task or domain within ONAP. The main modeling language is TOSCA. BPMN, DROOLS, OpenStack Heat, UML, XACML and YANG are also used.

TOSCA

TOSCA is a cloud-centric, human-readable modeling language. TOSCA supports both declarative and imperative paradigms. TOSCA is used in a wide range of use cases, including NFV, and defines cloud-based services, their components, relationships and the workflows that manage them. TOSCA also allows definitions of workflow triggers. TOSCA defines profiles. The TOSCA simple profile for NFV is used to define an NFV centric data model.

```
homestead_host:
   type: clearwater.nodes.MonitoredServer
   Capabilities:
      Scalable:
       Properties:
          min_instances: 1
   Relationships:
      - target: base_security_group
        type:
cloudify.openstack.server_connected_to_security_group
      - target: homestead_security_group
        type:
cloudify.openstack.server_connected_to_security_group

Homestead:
   type: clearwater.nodes.homestead
   Properties:
     private_domain: clearwater.local
     release: { get_input: release }
   Relationships:
     - type: cloudify.relationships.contained_in
       target: homestead_host
     - type: app_connected_to_bind
       target: bind
```

Sample TOSCA code

YANG

YANG is a network-centric modeling language. It was initially tied to NETCONF, a network equipment configuration protocol. It is human-readable and models state data and the configuration of network elements. It supports hierarchical config data models, reusable types and groupings, RPC calls via NETCONF and NETCONF notifications.

```
module acme-system {
    namespace "http://acme.example.com/system";
    prefix "acme";
    organization "ACME Inc.";
    contact "joe@acme.example.com";
```

```
    description
        "The module for entities implementing ACME
system.";
    revision 2007-11-05 {description "Initial revision.";}
    container system {
        leaf host-name {
            type string;
            description "Hostname for this system";
        }
        leaf-list domain-search {
            type string;
            description "List of domain names to search";
        }
        list interface {
            key "name";
            description "List of interfaces in the
system";
            leaf name {type string;}
            leaf type {type string;}
            leaf mtu {type int32;}
        }
    }
}
```

Sample YANG model

Other Modeling Languages

Language	Description	Primary Functions
BPMN	Business Process Model and Notation (BPMN) defines the steps in a business process/workflow (or network service). It uses graphical notations and is used for defining workflows attached to a network service.	Orchestration workflow
XACML/ DROOLS	XACML is used to define simple policies while DROOLS is used for	Policy

	more complex ones.	
OpenStack Heat	OpenStack Heat is a declarative model for orchestrating OpenStack resources and managing their lifecycle.	VNF onboarding
UML	General purpose, developmental modeling language. It is useful for information models.	Information modeling

Cloud Native Design Principles

Cloud native is a modern software architecture with the following key principles:

- Virtualized infrastructure resources
- Shared standard cloud platform
- Microservices/decoupled single capability modules with open APIs
- Separate of stateful and stateless microservices; common data layer and information models
- Scale-out instead of scale-up; availability will become part of the application instead of infrastructure, failure of individual VNFs is expected and handled with scale-out HA
- Backwards compatible
- Secure, reusable, antifragile

Cloud native applies to all components of the NFV stack, including VNFs and ONAP itself.

DevOps

DevOps (certainly an overused term) to me means continuously delivering small incremental changes to software applications via automation, as opposed to large multi-month changes. DevOps delivers innovations to users rapidly and, contrary to intuition, a DevOps approach is less risky than large step-function changes. DevOps requires both a technology and culture change. On the technology side, DevOps requires continuous integration (CI), continuous testing, continuous delivery (CD) and continuous monitoring. On the culture side, it requires that the silo between development and operations be broken down. Similarly, silos between server, storage and networking groups need to be broken down as well with the move to virtualization.

DevOps also applies to all software components of the NFV stack including VNFs and ONAP.

Other Architectural Principles

Additional ONAP architectural principles include:

- The use of pluggable modules
- Centralized design studio
- Policy driven with closed loop automation (and open loop ticket creation)
- Real-time in terms of minutes, seconds, and milliseconds
- Self-service for end-users, operators and designers
- Multi-tenant (allowing multiple network services to share resources) and secure
- Complete automation
- Multi Cloud (automating services running on different cloud infrastructures)

Day in the Life of ONAP

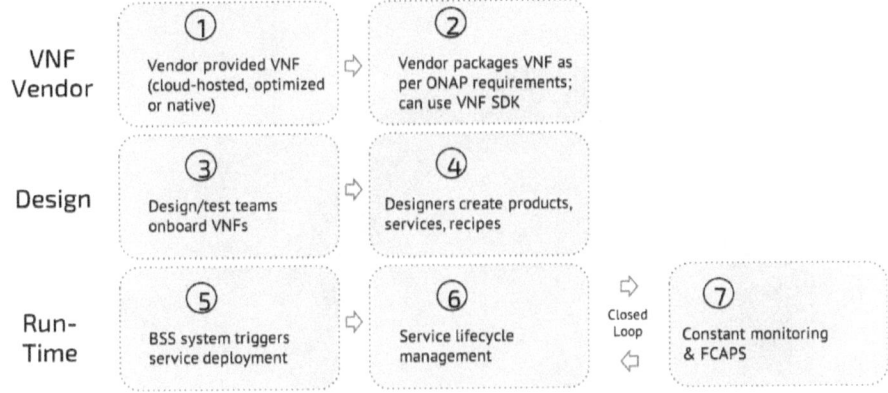

ONAP Sequence of Steps

Three types of organizations/roles are involved with ONAP. The first entity to get involved is a VNF vendor. The VNF vendor writes a cloud-native VNF that matches ONAP specifications and then packages it as per ONAP requirements. Next, the design and test teams onboard the VNF(s), create network services and recipes. Finally, the operations team is involved during run time, when these network services are running. By maximizing the number of closed-loop automation recipes created by the design team, the operations team gets freed up for a plan/build mindset instead of a break/fix mentality.

ONAP Releases

ONAP is on a 6-month release cycle. The third release, code-named Casablanca, was in December 2018. Contributions by organization are shown below.

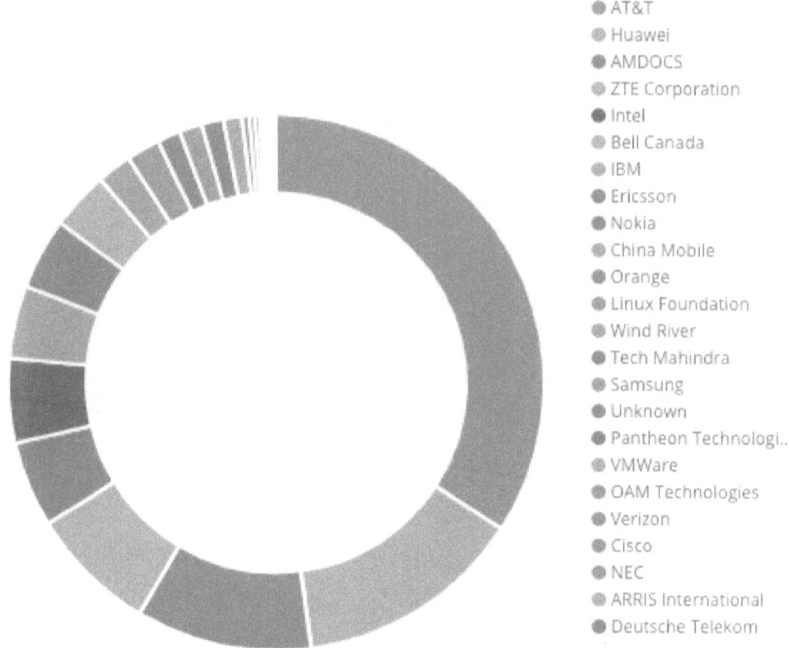

Casablanca Contributions

The next three releases, slated for 2019/20, are Dublin, El Alto, and Frankfurt.

Conclusion

ONAP is a MANO++ open source project from the Linux Foundation resulting from the merger of OpenECOMP software and the Open-O MANO project. ONAP will change how CSPs build and operate their networks, and it therefore includes several new and exciting architectural concepts. ONAP will harmonize the way VNFs and network services are deployed, simplifying the interaction between CSPs and VNF vendors. Finally, ONAP has a clean separation of the roles and activities of VNF vendors, designers and operators. In the next chapter, we will look at the ONAP architecture.

3

ONAP ARCHITECTURE

Introduction

In this chapter, we will we will look at ONAP architecture, cover the two major systems (design time and run time) and discuss the three major interfaces: VNF onboarding, OSS/BSS and VIM/SDN Controller. We will also look at how the different ONAP modules work together as one and help you move from a break-fix mindset to a plan-build one. We will finish by revisiting how ONAP enables agility and taking a look at what it takes to implement ONAP because it is not a push-button exercise.

ONAP Architecture

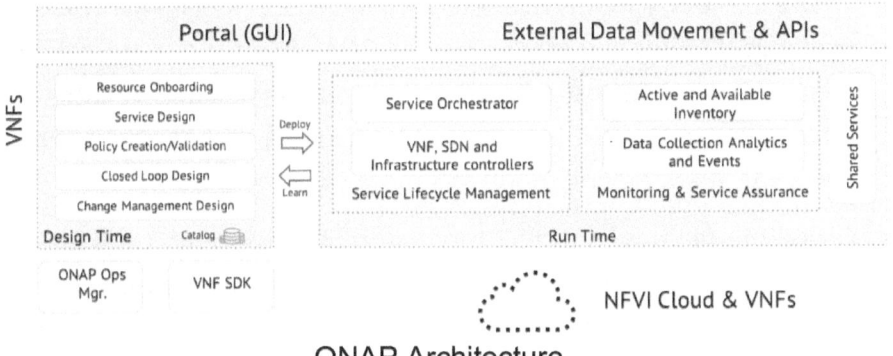

ONAP Architecture

ONAP architecture defines two major systems: design time and run time. This separation allows for clean delineation between design and operational roles. These modules are responsible for:

- Design time environment:
    - VNF onboarding/validation
    - Network service/SDN service design
    - Policy creation
    - Workflow design
    - Analytics application onboarding
    - DCAE workflow design
    - Event monitoring
- Run time environment:
    - Service orchestration
        - Service orchestration & lifecycle management
        - VNF controller (VNF orchestration & lifecycle management)
        - Infrastructure controller (interface to the VIM, SDN controller)
    - Monitoring and service assurance
        - Data collection, analytics and events
        - Storage of all active and available inventory

In addition to these two major modules, there is a GUI and an API interface. Next, a module called ONAP Operations Manager (OOM) manages the lifecycle of ONAP, since ONAP itself is a cloud-native application that needs to be orchestrated and managed. Another module called the VNF SDK helps VNF vendors package their VNF as per ONAP requirements.

ONAP interfaces with three major external subsystems. On the northbound interface (the interface with higher-level layers of software), ONAP talks to OSS, BSS, big data analytics and E-services applications. On the southbound interface (the interface with lower-level layers of software), ONAP communicates with the VIM, the NFVI and the SDN controller jointly constituting the NFV cloud. ONAP also onboards VNFs and analytics applications.

ONAP Design Time Environment

The ONAP design time environment supports four activities:

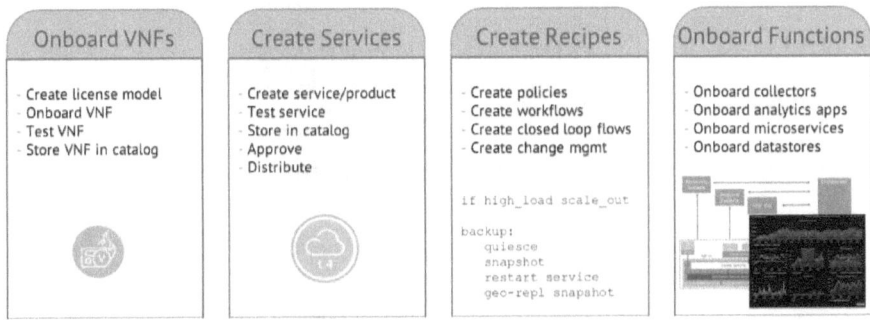

ONAP Design Time Activities

These activities are supported by two graphical tools:

- Service Design & Creation (SDC)
- Closed/Control Loop Automation Management Platform (CLAMP)

ONAP Run Time Environment

The ONAP runtime environment consists of several software modules:

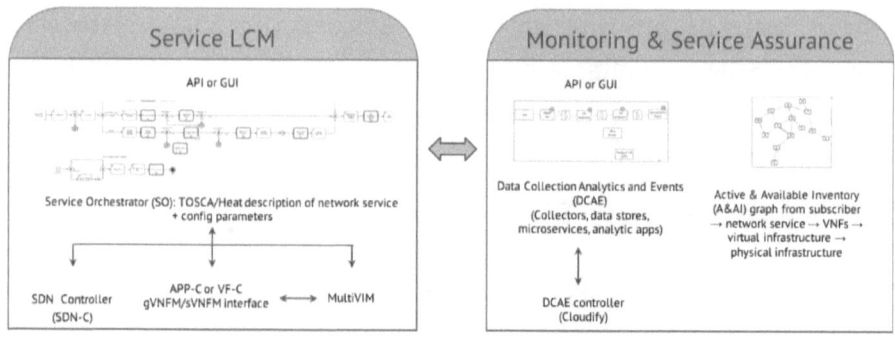

ONAP Run Time Activities

Service orchestration and lifecycle management consists of:

- Service Orchestrator
- SDN-C — SDN controller
- APP-C (Application Controller) & VF-C (Virtual Function Controller) — VNF managers, interface to external sVNFM
- MultiVIM — Infrastructure/VIM manager (controller)

Monitoring and service assurance consists of:

- A&AI — module that tracks active and available inventory
- DCAE — Data collection analytics and events; just like SO, it has its set of controllers. DCAE also has its own controller for spinning up virtual infrastructure and underlying components

We will look at these modules in much more detail in the next chapter.

VNF Onboarding Process

Onboarding VNFs onto ONAP is an important task. It occurs at three levels.

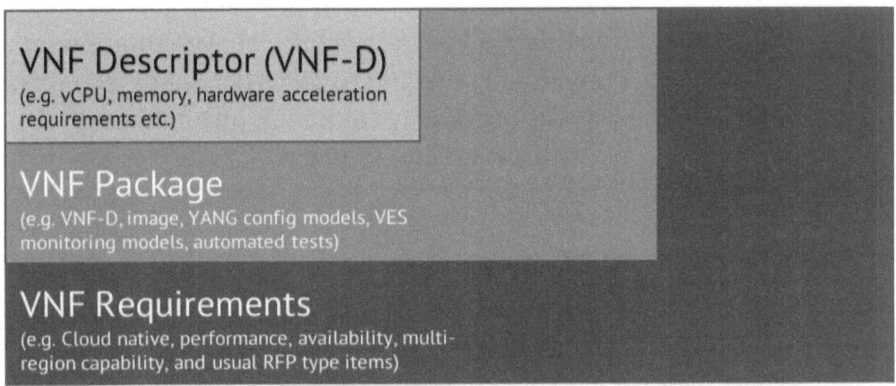

VNF Onboarding Process

At the highest level, there are VNF requirements outlined by the ONAP project. Next, ONAP describes the overall ONAP package. Finally, the VNF descriptor, part of the VNF package, is also described in detail. The VNF descriptor (VNFD) may be provided in OpenStack Heat or TOSCA formats.

While we are focusing on VNF onboarding, DCAE artifacts (analytics apps, microservices, data stores, etc.) are also onboarded through SDC.

VNF Onboarding Projects

Given the importance and complexity of VNF onboarding, ONAP has three projects dedicated to this.

Project	Details
VNF requirements	A documentation project that produces specifications for VNF vendors to follow.
VNF SDK	Software tooling that VNF vendors can use to ease the creation of a VNF package.
VNF Validation	An environment and set of processes that

> validate a VNF against the above requirements, eventually leading to an ONAP validated label. It was previously called ICE (Incubation & Certification Environment).

ONAP OSS/BSS Interface

ONAP provides API access to just about every major component. These APIs, which over time will be aligned with MEF and TMForum, can be used for OSS/BSS integration. The types of service supported through these interfaces include:

OSS

- Capacity management
- Handling major failures (e.g., at a region level)
- Providing KPIs

BSS

- Interface to Service Orchestrator (SO) to invoke network service instantiation along with run time configurations and query the status of services
- Billing information, user events, customer SLA management from DCAE/data lake
- Customer information management (CIM) via A&AI
- SLA (availability, performance) enforcement, etc. via Policy

ONAP as a Whole

ONAP Components Working Together

The various ONAP components work together as one through the above steps. The level of automation that can be achieved through this flow allows the operational staff to move from a break-fix mindset to a plan-build one.

ONAP Enables Agility

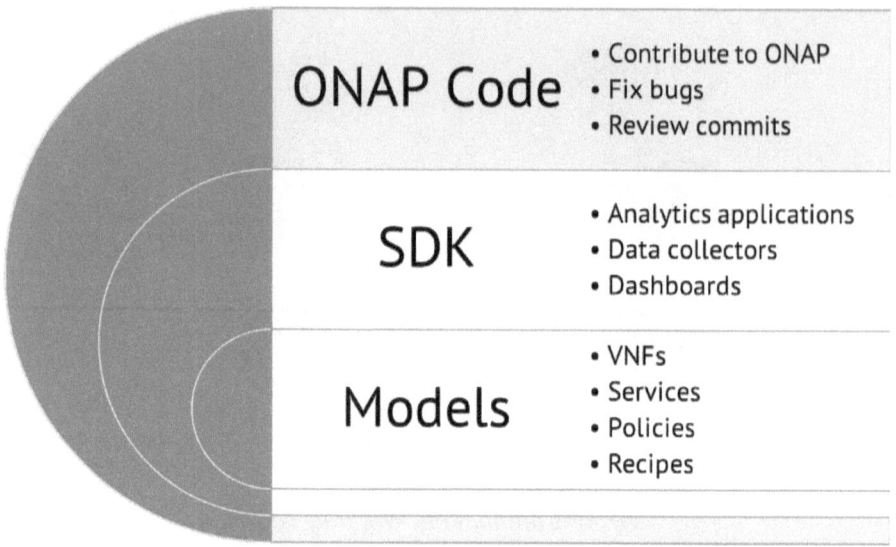

ONAP Enables Agility

ONAP brings agility to a CSP in three ways:

- A model driven approach allows automation without coding. VNFS, network services, policies, recipes, etc. can all be modified and changed by simply writing and editing models.
- Not everything can be changed through models. For certain items, ONAP provides SDKs for easy integration and customization. For example, analytic applications, data collectors and dashboards can be written by CSPs or vendors using different ONAP SDKs. While this approach does not eliminate the need for programmers, it does mean that changes can be made without changing the ONAP codebase.
- If a CSP or vendor needs an even deeper level of customization within the ONAP codebase, they can simply contribute those

changes because ONAP is 100% open source. Through this mechanism, CSPs can gain a level of influence on the software not possible with proprietary products.

ONAP Implementation

The implementation of ONAP by a CSP requires integration with different aspects of vendor products and existing software systems. The implementation steps may be summarized as follows:

1. Onboard VNFs
2. Design network services
3. Write workflows
4. Create policies
5. Onboard or write data collectors
6. Build analytic applications
7. Design closed loop automation flows
8. Interface to OSS/BSS systems
9. Install the NFVI/VIM layers (possibly VNFMs as well)

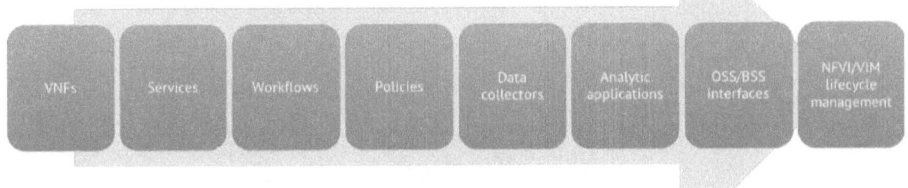

ONAP Implementation Steps

Conclusion

ONAP is cleanly structured into design time and run time modules. The design time environment is unified, guided, role-based and extensible. The run time environment supports service orchestration/lifecycle management and monitoring/service assurance. The ONAP community has also invested in the three main interfaces in terms of VNF onboarding, OSS/BSS interfaces and NFVI/VIM/SDN Controllers. ONAP components work smoothly together to enable operations and move from a break-fix to plan-build mentality. However, in order to generate business value, ONAP needs to introduce an implementation phase.

4

ONAP DEEPDIVE

Introduction

In this chapter, we will take an in-depth view of the various sub-projects[2] that make up ONAP. The Casablanca release has 30 official sub-projects. We will discuss each of the major projects and superficially cover the remaining projects. Our discussion will also give us a look into the project organization. It balances autonomy with governance. While each project can largely make its own decisions, the technical steering committee, or TSC, and a set of subcommittees (such as architecture, use case and security) drives consistency and requirements across projects. The code created by these projects is completely open, as is the process.

---

[2] Technically these are sub-projects since ONAP is a Linux Foundation project. However, we use the terms project and sub-project interchangeably.

Official ONAP Projects

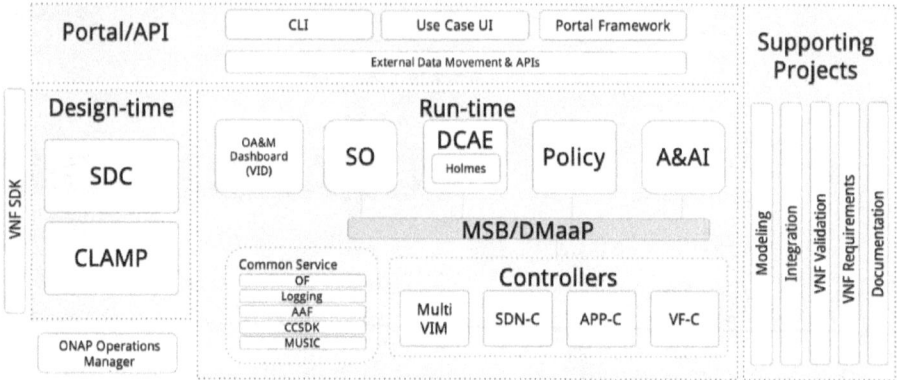

ONAP Detailed Architecture

We will discuss some of the major software projects below.

ONAP Lifecycle Management

ONAP itself is a cloud native application that needs orchestration and lifecycle management. The list of management tasks ranges from:

- Initial deployment
- Configuration
- Configuration changes
- Scale-out
- Self-healing
- Updates/upgrades

This is accomplished by using a containerized version of ONAP with Kubernetes (k8s) and Helm.

ONAP Design Time: SDC

The Service Design & Creation project is a unified tool for design-time activities such as:

- Onboarding VNFs
- Creating services
- Creating policies
- Creating workflows including change management
- Onboarding data collectors
- Onboarding analytic apps
- Creating closed loop templates
- Testing, approving and distributing artifacts to run time

SDC Screenshot

SDC is catalog-guided, driven, extensible and role-based (roles such as design, test, admin, governance).

CLAMP

The CLAMP tool is used to configure policies/templates created in SDC to create closed-loop service assurance loops. The below figure shows a template being configured with the specific collector, threshold and string match; the template is then "attached" to a network service. CLAMP also includes an event monitoring dashboard that can be used for activities such as debugging a closed loop by monitoring DMaaP.

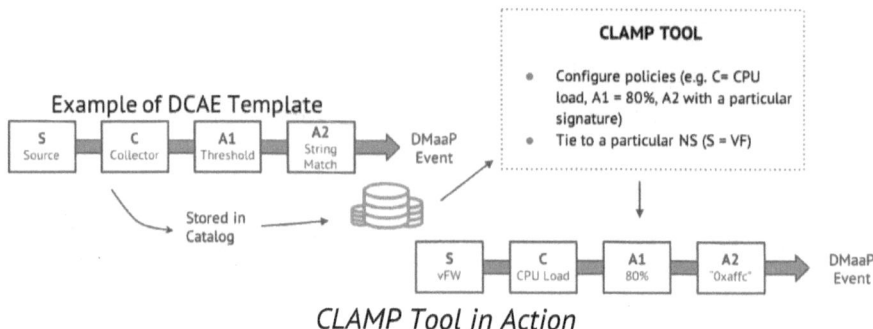

CLAMP Tool in Action

SO

The Service Orchestrator (SO) is used for real-time service creation or lifecycle management. It cuts days' or weeks' worth of effort to <10 minutes. It primarily interacts with controllers, SDC and A&AI. SO gets topologies and configurations through resource and recipe models via SDC. The SO is invoked through events or APIs from BSS or manually invoked through VID (Virtual Infrastructure Deployment). Run time configurations for network services are also provided when SO is invoked for either initial network service deployment or post-deployment configuration change. SO makes homing (where to place the workload) decisions, resource determination decisions and makes the triggers management actions via one of the four controllers at its disposal. SO also handles errors and rollbacks.

Service Type, Request Action, Variables (via APIs or VID)

⇕

SO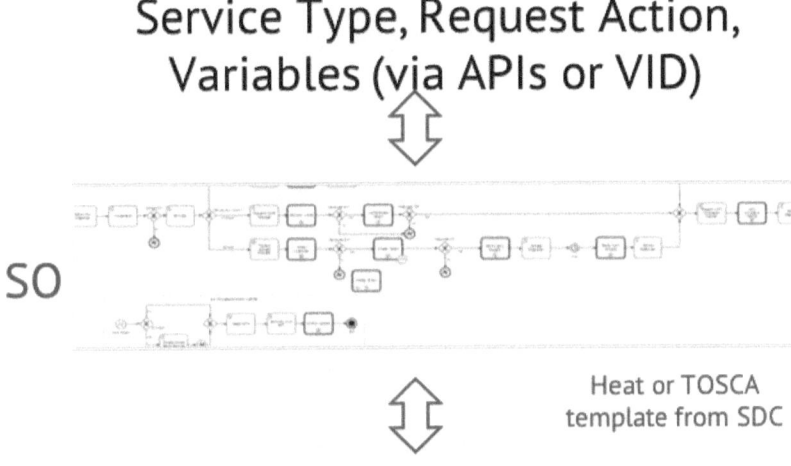

⇕ Heat or TOSCA template from SDC

A&AI, Respective controller

ONAP SO Interfaces

ONAP Run Time: Controllers

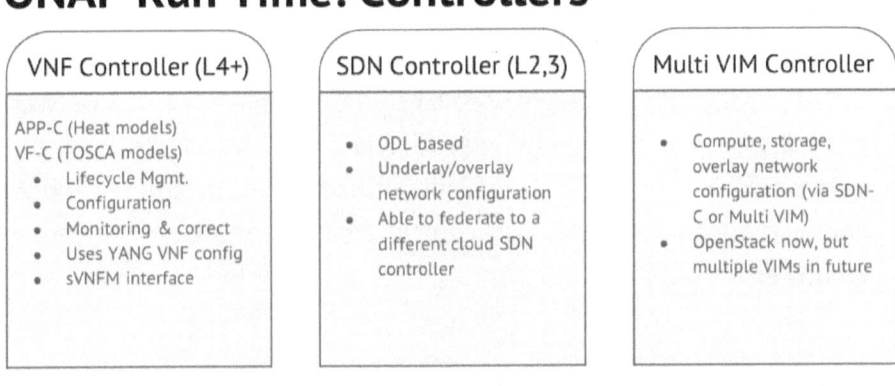

Controllers can be invoked by SO or directly based on events, VID or APIs

The four controllers are APP-C (Application Controller), VF-C (Virtual Function Controller), SDN-C (SDN Controller), and MultiVIM. The roles are shown in the above figure. The controllers act under the direction of SO or independently as triggered by events. As evidenced, there are two methods for service orchestration in ONAP; one leverages APP-C and the other VF-C.

APP-C

APP-C, or application controller, loosely maps to the functionality of a generic VNF manager (gVNFM). It performs lifecycle management of VNFs. Its focus is therefore L4-L7.

VF-C

VF-C, or virtual function controller, loosely maps to the functionality of an orchestrator and can interoperate with external specific VNF managers (sVNFM) through a driver or over the ETSI NFV-SOL003 standard interface. It accepts VNF descriptors in TOSCA format and manages initial orchestration and lifecycle management of network service and VNFs, directly or through the sVNFM.

SDN-C

SDN-C, or SDN controller, is used to configure layers 2 and 3 networks, both physical (underlay) and virtual (overlay). It can coexist with an external SDN controller; often, users utilize SDN-C for inter-cloud WAN connectivity and external SDN controllers for connectivity inside a cloud region.

MultiCloud (or MultiVIM)

The MultiCloud communicates with the VIM, which, in this case, is OpenStack. The term MultiVIM, for now, indicates that ONAP can connect to multiple OpenStack clouds that are from different vendors and/or different versions. There is also experimental support for other VIMs, e.g., Microsoft Azure and Kubernetes.

DCAE

The data collection analytics and events, or DCAE, module is used for closed-loop automation, trending, solving chronic problems, capacity planning, service assurance, reporting and so on. DCAE orchestrates data collectors, microservices, analytic applications and closed loops. To orchestrate these elements, the DCAE uses a controller called, for simplicity, the DCAE controller. Analytic applications may be written using a well-documented CDAP (Cask Data Application Platform) SDK. Given the model- and SDK-driven nature of DCAE, it offers a self-service interface to for developers, designers and operators. DCAE interacts with SDC, Policy and A&AI.

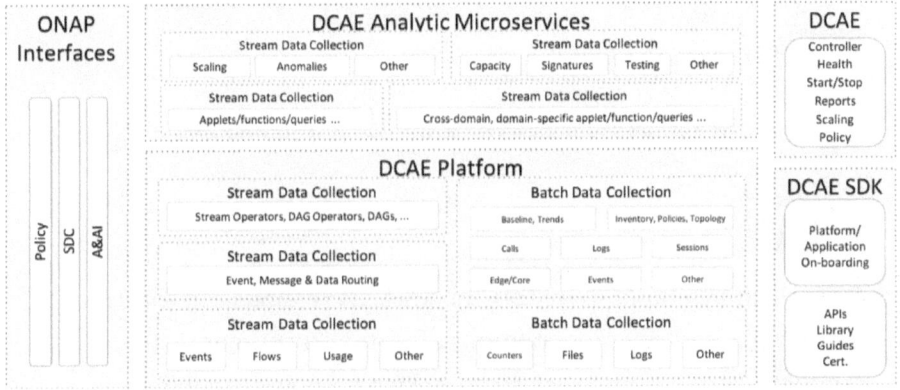

DCAE Architecture

The above architecture diagram shows the various elements of the DCAE platform, microservices, DCAE controller, SDK and interfaces.

Policy

The Policy software module guides the automated system without code, purely through models that allow for dynamic changes and a self-service approach for designers (i.e., designers do not have to go through software engineers or operators to make changes). Policies can guide a wide range of activities, including service instantiation, data collection, control loops, change management and so on. Simple policies may be specified in YAML (XACML), while complex policies can be described in DROOLS, all through SDC. There is basic policy conflict validation available, with more sophisticated validation to come in future releases.

```
if cpu_load > 80% && duration > 5 minutes {
      add 1 instance of VNF
}
```

```
if physical_memory < 15GB {
      migrate VNF
}
```

```
If memory_usage increases 5GB or more && duration < 10 seconds {
      Restart VNF
}
```

Examples of simple policies (pseudo-code, not in any modeling language)

A&AI

Active and available inventory, or A&AI, is responsible for dynamic, real-time inventory management of virtual resources. It tracks products, services and resources, both active and available. Additionally, A&AI tracks relationships by maintaining a graph database, e.g., subscriber → NS → VNFs → VMs → Compute/storage/networking nodes. A&AI APIs allow discovery, registration and auditing. Like other systems in ONAP, A&AI is model driven and extensible. As you can imagine, A&AI is the single source of truth in ONAP. It is a real-time system, and, along with audit capability, it is suitable for a CI/CD flow. A&AI can also maintain references to physical network functions and incorporates a sub-component called External System Registry (ESR). The ESR can register external systems like the VIM, sVNFMs, external SDN controllers etc. A&AI interacts with most ONAP components.

ONAP Portal

The ONAP portal provides a single pane of glass for design (e.g., SDC, CLAMP) and OA&M (e.g., VID, UUIs that are covered in the next chapter). It is also extensible via third-party apps. Third-party portal apps can choose different levels of integration: basic, target or extended integration. Portal apps are envisioned to be in 7 areas: design, ops planning, capacity planning, technology management, technology insertion, performance management and platform management. Everything that can be done through the GUI is also available through APIs. To interact with APIs in an easy manner, ONAP also provides a command line interface (CLI).

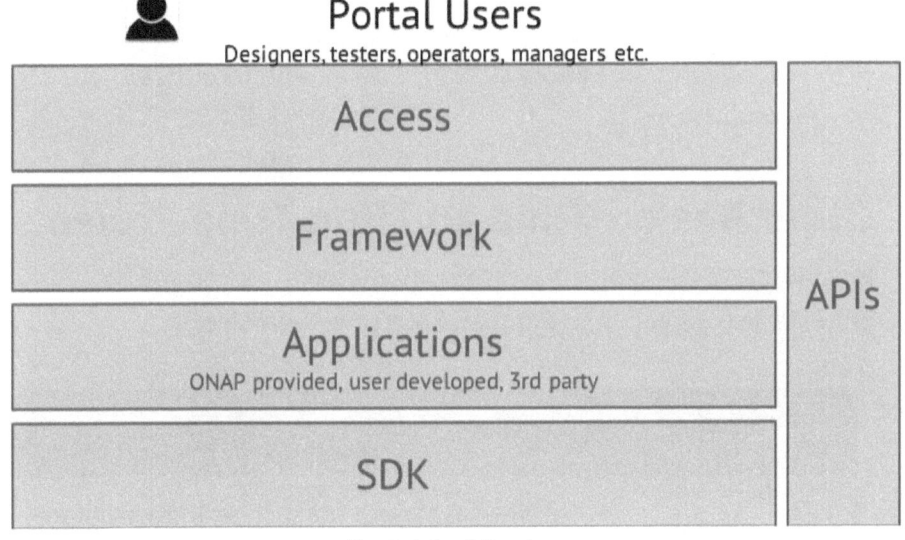

Portal Architecture

ONAP Northbound APIs

Everything that can be performed through the GUI or CLI can also be invoked through a northbound API. In fact, the API is a superset. The APIs are being aligned with TMForum APIs and, over time, MEF APIs. The relevant TMForum APIs are:

- Service Catalog (TMF633)
- Service Order (TMF641)
- Service Inventory (TMF638)

The relevant MEF APIs are:

- Allegro (CUS:SOF): Dynamic service control, Service state info, Service performance & quality, Service related alerts
- Legato (BUS:SOF): Service feasibility, Service configuration & activation, Usage events & metrics, Service performance & quality, Service policy
- Interlude (SOF:SOF): Dynamic service control, Service parameter configuration, Service state info, Service performance info, Service problem alerts

Other Design Time and Run Time Projects

In addition to the projects we have discussed, these additional four projects also pertain to design and run time activities.

Project	Goal
VID	<ul><li>O&M dashboard for service/infrastructure deployment, configuration, change management</li><li>Essentially manual interface to SO</li></ul>

UUI	• Special dashboard for design, lifecycle management and monitoring of specific network services e.g. ONAP VoLTE and CCVPN blueprints
Holmes	• DCAE app (or standalone) for alarm correlation and analysis for RCA • Works in conjunction with Policy by reducing the alarm load on Policy
VNF SDK	• Create tools/frameworks to expand the ONAP compatible VNF ecosystem: package validation, lifecycle test, functional test, CI/CD toolchains and so on • Works closely with VNF requirements and VNF Validation projects • Will enable a VNF marketplace between vendors and users

ONAP Common Services

ONAP provides a number of common services used by different projects:

Project	Goal
AAF	• Centralized Authentication and Authorization Framework for ONAP components • Fine-grained authorization to enable rapid deployment of new components

CCSDK	- Common reusable code that can be used across multiple controller projects; thus the name Common Controller SDK
DMaaP	- Data Movement as a Platform responsible for transporting and processing data within ONAP (intra/inter-regions) - Data filtering, data transport, data processing
Logging	- Improving log generation consistency across ONAP components, documentation - Reference ELK stack (Elasticsearch-Logstash-Kibana) for log processing/visualization - A new Post Orchestration Model Based Audit (POMBA) initiative allows audits between design and ops
MSB	- Microservices Bus (actually a mesh) provides a common infrastructure for ONAP services, e.g., registration, discovery, health check, API gateway, API portal, client SDK - MSB also integrates the Istio service mesh
MUSIC	- Standard services to distribute applications across multiple clouds to get 5x9s availability with <=3x9s of infra
OF	- Optimization Framework to assist other components - HAS (Homing and Allocation Service) helps SO/controllers choose the right region/cloud - CMSO (Change Management Scheduling

	Optimizer) allows designers to choose appropriate schedules for their change management apps

ONAP Supporting Projects

Finally, ONAP has several projects that are not part of the mainstream ONAP software, yet provide a critical role in a number of areas:

Project	Goal
Documentation	Create and maintain documentation for ONAP releases for various audiences: platform developer, admin, designer, tester, VNF developer, service provider, etc.
Integration	Tools, frameworks and pipelines for CI/CD, continuous testing, open labs. Also responsible for the Robot tool that is used for automated testing and emulation (e.g., BSS emulation).
Modeling	Unified data models for A&AI and SDC: resource models, service models, deployment/lifecycle models.
VNF Requirements	Comprehensive set of requirements for a VNF vendor: VNF package, VNFD, monitoring, configuration, testing, documentation and so on.
VNF Validation	Self-certification program for VNF vendors based on requirements review and testing; previously

	called ICE.

ONAP Project Structure

ONAP is organized as follows:

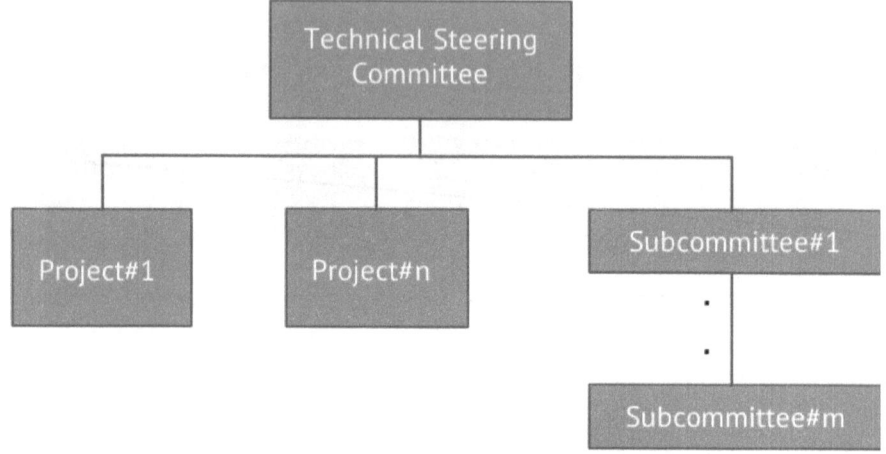

ONAP High Level Project Structure

The Technical Steering Committee, or TSC, manages the overall direction and provides oversight and governance of all underlying activity. The activities, broadly speaking, comprise projects and subcommittees. Projects target a specific need or software component. Subcommittees tackle concerns that affect multiple projects.

Anyone can propose a project. If approved, a project has a Project Technical Lead (PTL), committers and contributors. Anyone can become a contributor. Committer and PTL positions are merit-based and are voted on by the project itself. All project activities, meetings, meeting minutes, JIRA backlog, etc., are open.

ONAP Releases

The ONAP community releases a new version of the software roughly every six months. The release schedule is as follows:

Release	Date	Key Features
Amsterdam	Nov-17	• Initial integration of OpenECOMP & Open-O • 29 projects showing end-to-end orchestration, analytics and policy-based, real-time closed-loop automation
Beijing	Jun-18	• Northbound interface (NBI) alignment with TMForum • S3P: stability, security scalability, performance • Change management with in-place upgrades • Manually triggered scaling of VNFs • Containerized DCAE for OOM • 2 new projects: MUSIC, Benchmark
Casablanca	Dec-18	• Start of 2 new blueprints: CCVPN and 5G • 3 new dashboards • New LCM functions, PNF support • Expanded hardware platform awareness • Continued S3P (scalability, stability, security, performance)
Dublin	Mid-19	• First demo of 5G • Additional dashboards

El Alto	End-19	• To be determined
Frankfurt	Mid-20	• To be determined

Conclusion

ONAP is a vibrant project with subprojects in the categories of ONAP lifecycle management, design time, run time, Portal/API, common services and supporting projects. Each subproject has a project technical lead (PTL) and a dedicated set of committers and contributors. All projects are open, and anybody can contribute.

5

ONAP BLUEPRINTS & GETTING INVOLVED

Introduction

In this chapter, we will look at the three official ONAP demos that were part of the Amsterdam release. These are also called blueprints by the community. The three official blueprints are virtual firewall, residential virtual customer premise equipment, or vCPE ,and voice-over-LTE, or VoLTE. These blueprints serve multiple purposes. First, they give a clear indication of what is possible with ONAP. Next, they serve as "how to guides," where you can copy the best practices and techniques to implement your own POCs. Finally, these blueprints provide clear priorities to the ONAP community, in terms of what features and bugs to work on.

vFW Blueprint

The virtual firewall (vFW) blueprint is a rather simple one. It is meant to illustrate how ONAP works and to verify that an ONAP installation has completed properly. The entire network service consists of three VNF: a packet generator, a virtual firewall and a packet sink (sinc). The packet generator is onboarded as a separate VNF, and the virtual firewall and sink are onboarded as a compound VNF. The firewall blueprint has one metric that is being collected: the number of packets passing through the firewall in a 10 second period. There is a policy that checks if this number is greater than 700 or less than 300. If so, an APP-C workflow configures the packet generator to output 500 packets per 10 seconds. In this manner, all aspects of ONAP are demonstrated and exercised.

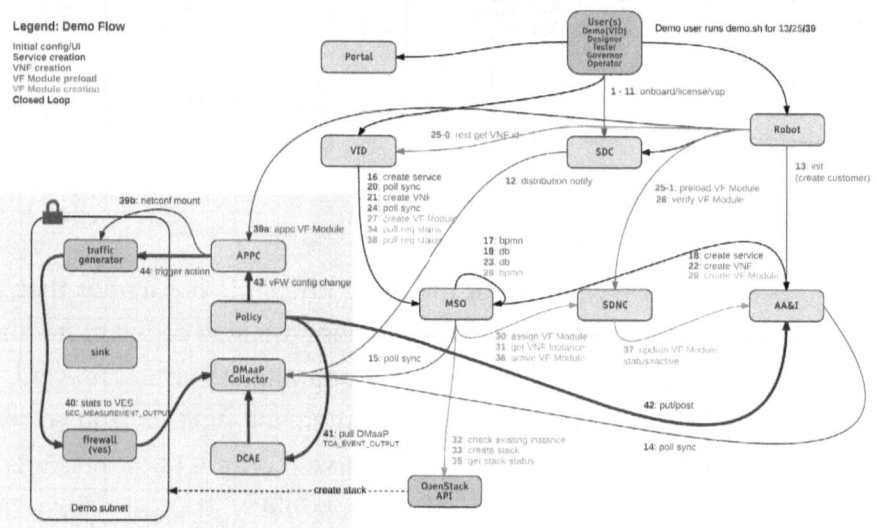

vFW Demo Flow

The demo goes through 40+ steps that include:
- VNF Onboarding
- Service and policy creation
- Service and policy distribution
- Service deployment using VID
- Traffic generation
- Closed loop automation

Residential vCPE Blueprint

This blueprint takes a residential gateway and splits it into two parts. The first part is a bridged residential gateway (BRG), basically a simple bridge, that resides at a customer's residence. The second part is a virtual gateway (vG) that resides in the CSP's central office or NFV cloud. All hardware functions previously present in a residential gateway are now virtualized and managed in the vG. If the idea was simply to move the functions into the cloud, the concept would not be that exciting. The real potential of this demo is to show CSPs how they can introduce new services such as AR/VR, tactile internet, premium HD/360° video and targeted ad-insertion to generate new revenue.

vCPE Technical Walkthrough

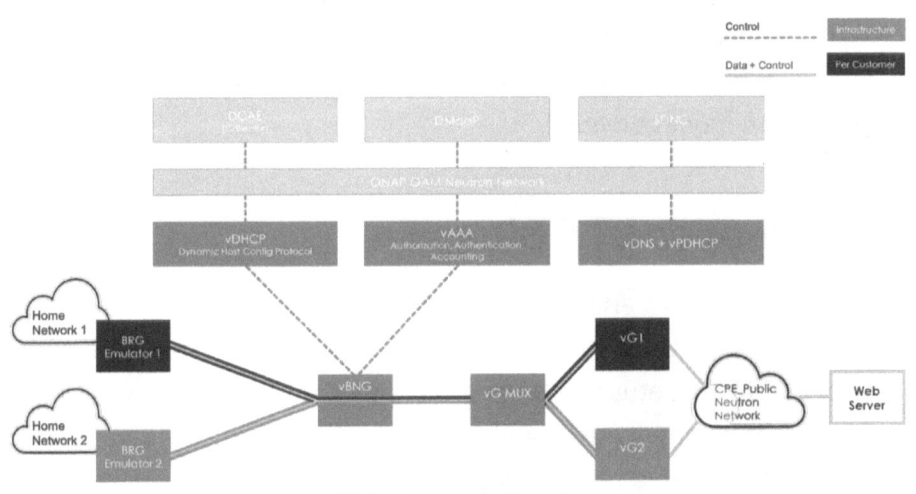

vCPE Network Service

The vCPE blueprint has the following characteristics:
- All VNFs are open source, though some are approximations not suitable for a production environment
- Several VNFs have been optimized using FD.io/VPP
- The demo uses APP-C, i.e., a gVNFM

- The overall demo consists of three constituent network services and three SDN services (6 services total)

The steps used to create the blueprint are:
- Register VIM with ESR
- Onboard VNFs
- Create services
- Create VNF placement policies for hardware platform awareness (HPA)
- Instantiate service through a BSS emulator
- Monitor service
- Closed loop automation
- In-place VNF upgrade
- Service termination

The vCPE blueprint also demonstrates change management and hardware platform awareness (HPA) using two approaches: policy-driven and model driven.

Voice-over-LTE Blueprint

CSPs can cut costs and add new services by moving to an all-IP network that includes voice-over-IP. Users experience higher quality calls and faster setups. The underlying technology, voice-over-LTE (VoLTE), is a combination of virtual IP multimedia subsystem (vIMS) and virtual evolved packet core (vEPC). The blueprint shows how ONAP orchestrates, monitors and manages a VoLTE service. The demo shows both vIMS and vEPC services straddling two clouds (representing edge and core).

ONAP Demystified

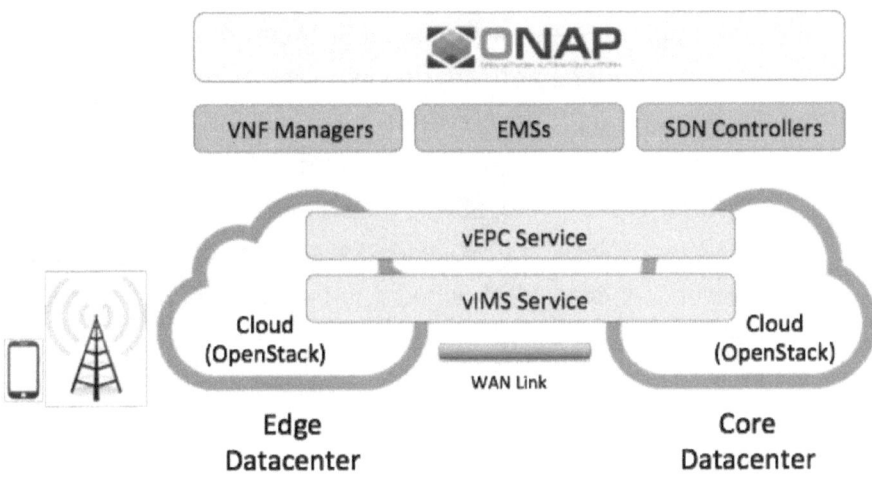

VoLTE Service

VoLTE Technical Walkthrough

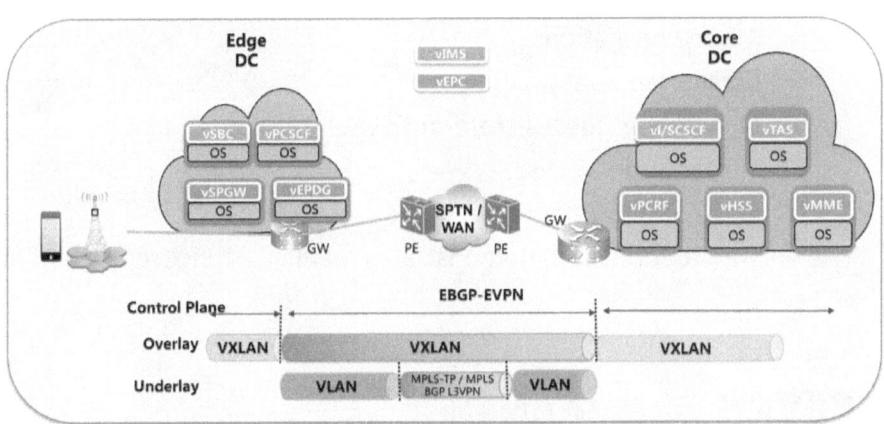

VoLTE Network Service

The VoLTE blueprint has the following characteristics:
- All VNFs are proprietary, production-grade and commercial provided by Huawei, Nokia, ZTE
- Since the VNFs are complex, they come with their own sVNFMs and element management systems (EMSs)

- The blueprint also uses external SDN controllers (Huawei, ZTE), a proprietary provider edge router (Huawei), and a DC/WAN/SPTN controller (ZTE)
- Appropriate drivers/interfaces were developed for VF-C (sVNFM), DCAE (EMS), SDN-C (SDN controllers)
- The overall VoLTE network service consists of two constituent network services (vIMS, vEPC) and two SDN services (underlay, overlay)

The steps used to create the blueprint are:
- Onboard VNFs
- vEPC/vIMS service design
- underlay/overlay VPN template import (service design)
- VoLTE service design
- Instantiate service through UUI
- Alarm correlation
- Closed-loop automation
- Manually triggered scale-out/scale-in
- Termination

The VoLTE blueprint also demonstrates manually triggered scaling of VNFs.

CCVPN Blueprint

CSPs see a strong demand for high-bandwidth, flat, high-speed OTN (Optical Transport Networks) across carrier networks. They also want to provide a high-speed, flexible, and intelligent service for high-value customers, and an instant and flexible VPN service for SMB companies.

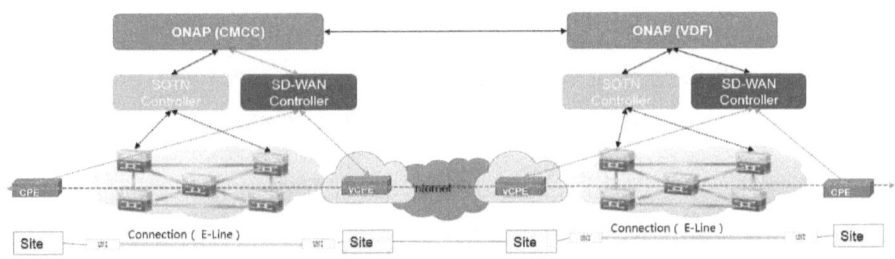

CCVPN Service

The CCVPN (Cross Domain and Cross Layer VPN) blueprint is a combination of SOTN (Super high-speed Optical Transport Network) and ONAP, which takes advantage of the orchestration ability of ONAP, to realize a unified management and scheduling of resource and services. It achieves cross-domain orchestration and ONAP peering across service providers. ONAP supports the CCVPN use case with several key components: SO, VF-C, SDN-C, Policy, Holmes, and DCAE. In this blueprint, SO is responsible for CCVPN end-to-end service orchestration working in collaboration with VF-C and SDN-C. SDN-C establishes network connectivity, then the VF-C component completes the Network Services and VNF lifecycle management. ONAP peering across CSPs uses east-west API which is being aligned with the MEF Interlude API. The key innovations in this use case are physical network discovery and modeling, cross-domain orchestration across multiple physical networks, cross operator end-to-end service provisioning and close-loop reroute for cross-domain service.

CCVPN Technical Walkthrough

The CCVPN blueprint involves the following tasks:

- Service and resource on-boarding and design
- Service deployment and configuration with ONAP⇔ONAP communication
- Self-service adaptation (closed loop showing bandwidth on demand, subsequent release)

- Auto-scaling based on fault and performance (subsequent release)
- Fault detection and auto-healing (subsequent release)
- Data correlation and analytics (subsequent release)
- Service termination

During the onboarding and design phase, 4 services are created using SDC:

- SOTN VPN Infrastructure Service
- SD-WAN VPN Infrastructure Service
- Site DC Service
- Site Enterprise Service

5G Blueprint

The 5G blueprint is a multi-release effort, with Casablanca introducing first set of capabilities around PNF integration, edge automation, real-time analytics, network slicing, data modeling, homing, scaling, and network optimization. The combination of eMBB that promises peak data rates of 20 Mbps, uRLLC that guarantees sub millisecond response times and MMTC that can support 0.92 devices per sq. ft. brings with it some unique requirements. First, ONAP needs to support network services that include PNFs in addition to VNFs. Next ONAP needs to support edge cloud onboarding as network services will no longer be restricted to just large datacenters but will proliferate a large number of distributed edge locations. Finally, ONAP needs to collect real-time performance data for analytics and policy driven closed-loop automation. These requirements have led to several initiatives within ONAP to holistically address the 5G blueprint.

Getting Involved

Developers and users can both get involved with the ONAP project. Here are some suggestions:

Developers	Users
• Start with ONAP wiki • Join the community by creating an LF ID (no charge) • Read about project(s) • Join mailing list, calls • Read about developing on ONAP, download tools • Read about integrating with ONAP • Attend events	• Start with ONAP wiki • Join the community by creating an LF ID (no charge) • Read about project(s) • Join mailing list, calls • Deploy ONAP • Try vFW, vCPE use cases • Deploy PoC with all artifacts and interfaces developed • Attend events

Conclusion

In addition to building the software stack, the ONAP community creates demos. These demos show what is possible with ONAP and provide direction to the community in terms of how to prioritize its work. These demos are also useful to CSPs new to ONAP because they can be used to set up a demo/POC in their lab and extract value out of the software immediately.

Collectively, ONAP end user members support over 60% of mobile subscribers around the globe. This makes ONAP the de facto network automation platform. Why wait, join now!

6

ADDITIONAL INFORMATION

Resource	URL
Official ONAP site	onap.org
ONAP wiki	wiki.onap.org
ONAP YouTube channel	youtube.com/channel/UCmzybjwmY1te0FHxLFY-Uog
ONAP project meeting list (required LF ID)	wiki.onap.org/calendar/calendarPage.action?spaceKey=DW&calendarId=260b8a93-05dd-476a-ab6c-ce5f14d46a34
Linux Foundation ID	identity.linuxfoundation.org
ETSI NFV ISG	etsi.org/technologies-clusters/technologies/nfv

ABOUT THE AUTHOR

Amar Kapadia is an NFV specialist and co-founder at Aarna Networks, an open source NFV company providing products and services around the Linux Foundation OPNFV and ONAP projects.

Prior to Aarna, Amar was the NFV product marketing head at Mirantis. Before Mirantis, he was responsible for defining and launching EVault's public cloud storage service (acquired by Seagate), based on OpenStack Swift. In total, Amar has over 20 years of experience in storage, server, and I/O technologies through marketing and engineering leadership positions at Emulex, Philips, and HP.

Amar lives in San Jose, California with his wife and two sons. When not working, Amar enjoys skiing and hiking and being a boy scouts merit badge counselor, and a LaunchX mentor.

www.ingramcontent.com/pod-product-compliance
Lightning Source LLC
Chambersburg PA
CBHW030453220526
45464CB00006B/2521